Meet Phil the gerbil. He is
a gentle animal and he likes
to sing. He has a nice, large
cage. But Phil dreams of a
change.

The gerbil cage is open!
Phil runs out and goes away.
But where will Phil go? Will
he be in danger?

Phil races down to the
river. He sees a large barge.
Phil jumps onto the barge
and takes a trip!

Inside the barge there is
a huge stage!

Phil steps onto this large
stage. Once he's on stage, he
starts to sing.

"Who is this stranger?"
asks a gopher. "He sings like
a gem!"

"Sing more, sing more!"
the mice and gophers shout.
They shout that phrase after
each song.

5

Phil the gerbil is a big hit!

His photo appears in the paper.

In fact, a large photo of him

appears on the first page!

Phil the gerbil is a huge
star! His phone never stops
ringing! The change in his life
is so strange!

Wake up, Phil! It was just a strange dream! Phil the gerbil is glad. He still has his nice, large cage!

The End